L
I
F
E

V
I
E
W
S

Published by Creative Education
123 South Broad Street, Mankato, Minnesota 56001
Creative Education is an imprint of The Creative Company

Art direction by Rita Marshall; Production design by The Design Lab/Kathy Petelinsek
Photographs by KAC Productions (Peter Gottschling), Richard T. Nowitz, Tom Stack & Associates
(Dave Fleetham, Barbara Gerlach, John Gerlach, J. Lotter Gurling, Victoria Hurst, Thomas Kitchin,
Joe McDonald, Gary Milburn, Randy Morse, Tsado/NASA, Mark Newman, Brian Parker, Ed Robinson,
Mike Severns, John Shaw, Tom Stack, Greg Vaughn, Dave Watts)

Library of Congress Cataloging-in-Publication Data

Frahm, Randy.
Islands : living gems of the sea / by Randy Frahm.
p. cm. — (LifeViews)
ISBN 1-58341-027-9
1. Islands—Juvenile literature. [1. Islands. 2. Island ecology. 3. Ecology.] I. Title. II. Series.
GB471 .F73   2001
551.42—dc21
                                    00-045166

First Edition

2   4   6   8   9   7   5   3   1

LIVING GEMS OF THE SEA

ISLANDS

RANDY FRAHM

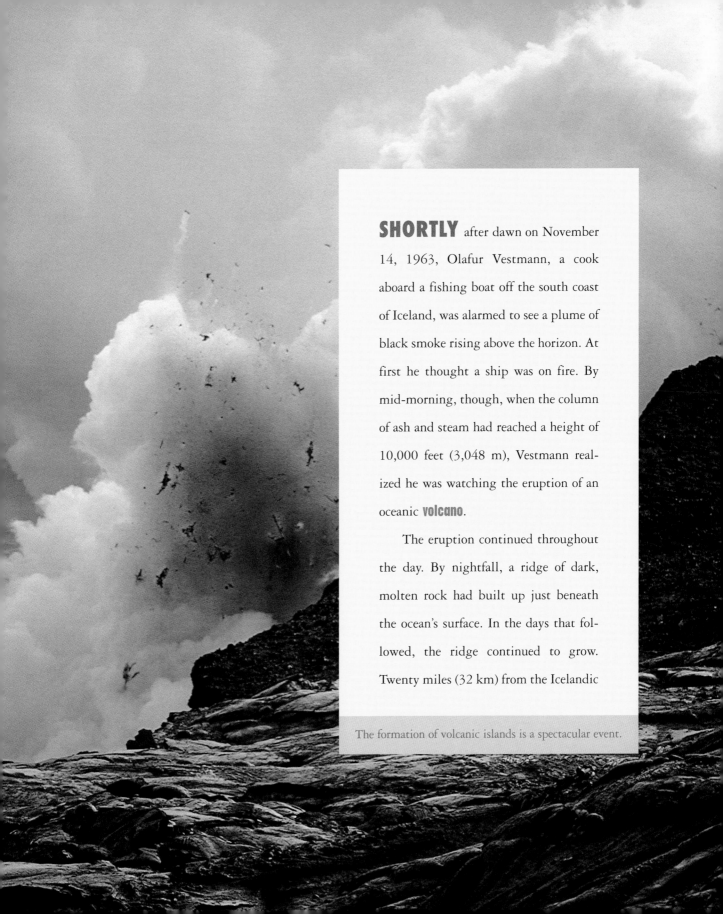

SHORTLY after dawn on November 14, 1963, Olafur Vestmann, a cook aboard a fishing boat off the south coast of Iceland, was alarmed to see a plume of black smoke rising above the horizon. At first he thought a ship was on fire. By mid-morning, though, when the column of ash and steam had reached a height of 10,000 feet (3,048 m), Vestmann realized he was watching the eruption of an oceanic **volcano**.

The eruption continued throughout the day. By nightfall, a ridge of dark, molten rock had built up just beneath the ocean's surface. In the days that followed, the ridge continued to grow. Twenty miles (32 km) from the Icelandic

The formation of volcanic islands is a spectacular event.

coast, an island was being born. It was named Surtsey, after Surtur, the Norse god of fire.

An island is defined as any body of land smaller than a continent and surrounded by water. Some, like Surtsey, are tiny outcroppings of **lava**, newly made and less than a mile (1.6 km) across. Others, like Greenland, contain some of the oldest rocks on Earth and are more than 1,000 miles (1,600 km) long. In the blustery North Atlantic, along the coast of Scotland, storm-lashed pinnacles of rock support only the hardiest of sea birds, while in the South Pacific, reef-fringed coral islands are home to a vast array of **wildlife**.

Most islands owe their existence to happenings deep within the earth. The surface of the earth is composed of a thin layer of rock called the crust. The crust floats on a layer of hot, partially molten rock called the mantle. According to the theory of **plate tectonics**, the crust is broken into a dozen large sections and many smaller ones called plates. These plates are

Most of the world's islands were formed by lava that flowed through cracks in the ocean floor, creating land masses that rose from the sea and hardened. These islands may be as small as Surtsey or as large as Iceland.

always moving, slowly but constantly, driven by the movement of hot rock in the earth's mantle.

The continents of North America and Europe sit on separate plates that are slowly moving away from each other. Where the plates are pulling apart—in the middle of the Atlantic Ocean—cracks develop in the earth's crust, and molten rock called lava flows up through them to form a prominent ridge on the **ocean floor**. Occasionally a ridge may extend above the surface of the ocean to form an island. Iceland is an example of an island formed in this way.

In other parts of the world, plates collide, and one plate is forced beneath the other. The edge of the lower plate melts, and the molten rock erupts, forming volcanoes. Some of these volcanoes extend above the ocean, creating long chains of islands called archipelagos, or **island arcs**. In the Pacific Ocean, island arcs run along a curved line that stretches from the Aleutian Islands near Alaska, through Japan and the Phillipines, to New Zealand.

The Galapagos Islands are an example of an archipelago formed by volcanic activity. This archipelago consists of 19 islands in all and includes many islets and rocks (opposite) that rise from the sea.

Scientists believe that the Hawaiian Islands were formed differently, not by colliding plates but by a single **plume** of molten material that rose from deep within the earth's mantle. Wherever the plume reached the surface of the earth's crust under the ocean, it created a hot spot. As the Pacific plate moved northwest, the plume melted patches of the crust and produced each of the islands in turn.

The age and size of each of the Hawaiian Islands seem to support this explanation. The largest island, Hawaii, is also the youngest and has the most active volcanoes. None of its rock is more than 500,000 years old. The other islands continue to the northwest for 1,200 miles (1,931 km) and become successively older—and smaller, too, since they have been exposed to the **erosional forces** of the ocean for a longer period of time. Kure, the uninhabited northernmost island, is more than 25 million years old.

Volcanoes are also the source of tropical islands known as atolls. An atoll is a circular or horseshoe-shaped **coral reef** that

Tropical islands such as the Galapagos Islands (top) and Palau (bottom) are renowned for the beauty of their landscapes and wildlife. The fertile volcanic rock that forms such islands supports lush vegetation.

surrounds a shallow body of water called a lagoon. A volcanic island once filled each area that is now a lagoon. Ocean waves brought tiny sea animals and plants to the edges of the island, where they established a home. Over time, the central volcano eroded away until it disappeared beneath the water's surface. Only the built-up edges of the fringing reef remained above the water.

A coral reef may be home to thousands of different creatures, including fish, shrimp, sponges, **sea anemones**, sea urchins, starfish, octopuses, and squids. The reef itself is formed by colonies of coral polyps—little animals related to sea anemones. Polyps secrete hard, cup-shaped skeletons of calcium carbonate into which they can withdraw if attacked. These skeletons are the newest layer of coral on a reef.

The polyps live in a mutually helpful relationship with a type of algae known as **zooxanthellae**. The zooxanthellae live in the tissues of the polyps, where they convert energy from the sun and carbon dioxide from the polyps into oxygen and carbohydrates. This conversion process, called **photosynthesis**,

Coral reefs are home to a magnificent array of sea creatures.

needs sufficient sunlight to occur, so coral reefs form only within the first 300 feet (91 m) below the ocean's surface.

Activity on the earth's surface can also create islands. Some islands consist of sediment, such as sand, silt, and gravel, that builds up along a shoreline. Streams and rivers wear away their banks and carry **sediment** to the shallow waters along seashores. Ocean waves and winds pile up the sediment into ridges and dunes, creating barrier islands. Barrier islands stretch 3,000 miles (4,827 km) from the Massachusetts coast to the Gulf of Mexico, forming the longest chain of islands in the world.

Other barrier islands were formed by **glaciers** that covered portions of the earth 10,000 to 15,000 years ago. As these massive sheets of ice and snow passed over an area, they pushed rocks, sand, silt, and clay in front of them. This debris formed a ridge called a **moraine**. When the earth warmed and the glaciers melted, some moraines were submerged, while others remained above the water, forming islands such as Long

Only the hardiest of plants and animals live on rocky islands such as James Island (top) off the coast of the northwestern United States. Many icebergs (bottom) are broken-off remnants of land-carving glaciers.

Island in New York and Nantucket Island in Massachusetts.

Other islands located near continents may be extensions of the continent itself. They appear offshore as small, isolated chunks of land. These islands are tall formations extending up from a huge underwater ledge—the continental shelf—that is the border of the continent.

When islands first form, they have no animal or plant life. Then, over time, plant seeds float across the sea or are carried by birds or the wind to an island, and **vegetation** begins to grow. Migratory birds may be blown off course by strong winds until, by chance, they come to an island. Some animals reach new islands by swimming. Reptiles and **mammals** too large to become airborne and unable to swim may be carried to islands on logs or other debris on which they have been marooned by floods or storms. Although many such creatures die on their long sea journey, a few survive and wash ashore.

As a result of this gradual, almost random process, many islands contain unique animals and plants. Hundreds or thousands of miles from the nearest

Many remote islands feature animals found nowhere else in the world. An example of this is the huge Galapagos tortoise (opposite), one of many unique animals native to the Galapagos Islands.

shore, new species **evolve** unaffected by outside competition. Plants and animals on the Galapagos Islands, located in the Pacific Ocean about 600 miles (965 km) from the coast of South America, are examples of this process. These islands are home to more than a dozen closely related species of finch that exist nowhere else, each having developed a different beak. Other unique Galapagos Islands inhabitants include a sunflower that grows as tall as a tree and a giant tortoise that weighs up to 600 pounds (272 kg).

Something similar appears to have happened on the island of Madagascar, which lies 250 miles (402 km) off the coast of East Africa. Madagascar has no ape or monkey populations, but it does have a variety of primitive **primates** called lemurs that exist nowhere else on Earth. One explanation for the development of these creatures is that the lemurs' ancestors evolved on the African mainland and then floated to Madagascar on rafts of vegetation. These creatures became isolated when Madagascar separated further from the African land mass. Once on the island, the primates evolved into many different species, ranging

The Coquerel's sifaka (top) is one of the largest lemur species found in the forests of Madagascar. Also making their homes on this tropical island are chameleons (left), mouse lemurs (middle), and geckos (right).

from the two-ounce (57 g) mouse lemur to the now-extinct 400-pound (182 kg) archaeoindrisa, a match for a large gorilla.

The ability of humans to find and inhabit remote islands is as remarkable as that of their animal counterparts. The ancient **Polynesians** of Fiji and the islands of Tonga and Samoa sailed thousands of miles over the Pacific Ocean, guiding themselves only by the stars.

Polynesians settled the Marquesas Islands by 300 A.D. and remote Easter Island a century later. They reached the Hawaiian Islands by 500 A.D. and New Zealand by 900 A.D. The first European **explorers** to arrive at these Pacific islands in the 1700s found societies thriving in a tropical paradise.

There is another side to the story of human **migration**, however. Human settlement on islands has often been accompanied by the widespread extinction of native island species of plants and animals. As people hunted and cleared the vegetation, they destroyed the natural habitat. They also introduced other species, such as rats and cattle, that

Many fish, plants, and crustaceans are native to the coral reefs and shallow waters off the coasts of tropical islands. Among these are starfish, crabs, shrimp, and brilliantly-colored tropical fish.

killed off the native island inhabitants through aggression or competition for the same limited food resources.

On Hawaii, researchers have documented the **extinction** of 50 species of native birds since the arrival of human settlers. New Zealand has a similar story. This land was once populated by various large, flightless birds. The most spectacular of these was the ostrich-like giant moa, which stood 10 feet (3 m) tall and weighed up to 500 pounds (227 kg). Other flightless birds included a giant coot and an enormous goose. **Fossils** show that these birds and their ancestors had lived for millions of years. Yet, within a few centuries of human arrival, all of these species were extinct.

Perhaps the worst example of **habitat** destruction took place on Easter Island. Lying 2,300 miles (3,701 km) off the coast of Chile, Easter Island is one of the most remote places on Earth. When the Dutch explorer Jakob Roggeveen visited the island in 1722, he found a bleak, treeless land inhabited by people constantly at war with one another. The ground was littered with enormous statues—stone heads up to 37 feet (11.3 m) tall and

Humans have had a great and irreversible impact on the Hawaiian Islands, including the islands of Maui (top) and Kahoolawe (bottom). As the human population of the islands grew, many native birds were wiped out.

weighing 80 tons (73 t) or more. Some statues remained in quarries, half-carved. It was as if one day the stone carvers simply put down their tools and stopped working.

Archaeologists have since reconstructed Easter Island's sad history. When the first people arrived on the island around 400 A.D., it was covered with dense forest. The settlers began cutting the forest for wood and clearing land to plant **crops**. The island's population grew too fast, and the forest was completely destroyed. The islanders no longer had timber for their houses and canoes or for the logs on which they moved their statues. **Soil** eroded from the naked land, and crop harvests declined. Eventually, the society collapsed.

The experience of Easter Island offers a lesson for all humanity. By misusing their island environment, the inhabitants destroyed their land and themselves. Islands have fascinated humans since ancient times, but only recently have we come to realize how **fragile** they are. As we continue to enjoy islands and learn from them, we must also take care to preserve and respect them.

Easter Island's famous giant statues are a reminder of its past.

MOVING OXYGEN

Many islands are like small worlds of their own, each harboring a unique array of wildlife. But this life is not concentrated only on the land above the sea; the ocean waters that surround these small land masses are home to many creatures as well. Coral reefs, for example, harbor some of the most colorful and diverse communities of life anywhere on Earth. Like animals that live on land, the creatures of the sea need oxygen to live. The following activity demonstrates the way that oxygen is dispersed through the ocean's water.

You Will Need

- A large glass or clear plastic container with a wide mouth
- Blue food coloring
- A measuring cup
- An ice cube tray
- A freezer

Water setup

1. The night before you conduct the experiment, use the food coloring to color some water a deep blue. Then freeze the water in the ice cube tray to produce blue ice cubes.
2. Pour about three cups (710 ml) of clear water into the container.
3. Add the blue ice cubes to the water in the container. Use as many cubes as necessary to form a solid layer of ice on the surface.

Observation

Watch what happens over the next few minutes. As the water at the top of the container gets cold from the ice, it gets heavier. This makes it sink to the bottom, carrying some of the blue color—which represents oxygen—with it. In much this way, water and oxygen are circulated through the ocean. Water sinks or rises, depending on its temperature. This movement creates patterns of water movement called currents, which help to stir the water, distributing oxygen and essential nutrients to all levels of the ocean. Though most of the oxygen in any given body of water remains near the surface, even creatures deep below get the oxygen they need to survive.

FLOATING ISLANDS OF ICE

Islands are masses of rock or sediment that rise from the earth's crust and jut out of the ocean. Many are able to support life—even entire ecosystems in some cases. Another kind of "island" exists in the waters near the northern and southern poles of the earth. These are massive, floating chunks of ice called icebergs.

Icebergs are not truly islands, of course, yet they can be quite similar in size and shape. One iceberg in Antarctica was measured at nearly 217 miles (350 km) long and 62 miles (100 km) wide. Such icebergs, which are broken-off sections of glaciers or polar ice sheets, appear huge upon the sea. But the portion that is visible above the surface is actually just a small part of the iceberg. About 85 percent of icebergs are underwater, making them even more colossal than they first appear.

To see this for yourself, try floating a little iceberg in your sink. First, fill a plastic container with water and set it in your freezer. When this water is frozen solid, add water to your sink until it is nearly full. Then set the block of ice in the water and watch what happens. Notice how most of the ice is submerged, with just a small portion rising above water.

LEARN MORE ABOUT ISLANDS

The Charles Darwin Foundation Inc.
100 Washington Street, Suite 232
Falls Church, VA 22046
http://www.galapagos.org/

Easter Island Home Page
(Online resource for information
    about Easter Island)
http://www.netaxs.com/~trance/
    rapanui.html

Galapagos Conservation Trust
5 Derby Street
London W1Y 7HD
United Kingdom
http://www.gct.org/

Komodo National Park—Komodo Island
P.O. Box 195
Ubud Bali 80571
Indonesia
http://www.komodo-national-park.com/

Sable Island Preservation Trust
P.O. Box 29028
Halifax, NS
Canada B3L 4T8
http://www.sabletrust.ns.ca/index.html

Virtual Galapagos
(Virtual expedition of the Galapagos
    Islands on the Internet)
http://terraquest.com/galapagos/

INDEX

Beautiful yet fragile, islands are small worlds of their own.